HOW DO YOU BUILD A
TIME
MACHINE?
AND OTHER
PUZZLES WITH SCIENCE

THIS IS A CARLTON BOOK

Published in 2015 by Carlton Books Limited
an imprint of the Carlton Publishing Group
20 Mortimer Street
London W1T 3JW

10 9 8 7 6 5 4 3 2 1

A catalogue record for this book is available from the British Library

ISBN 978 1 78097 669 3

Printed in Dubai

HOW DO YOU BUILD A
TIME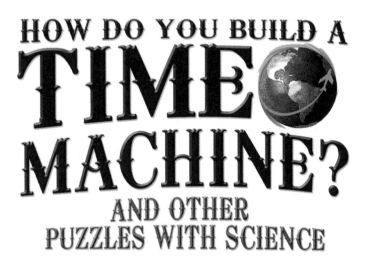
MACHINE?
AND OTHER
PUZZLES WITH SCIENCE

ERWIN BRECHER

CARLTON
BOOKS

About the Author

Erwin Brecher was born in Budapest and studied mathematics, physics, psychology and engineering in Vienna, Czechoslovakia and London. He wrote his first manuscript in 1992, which was published in 1994. Since then, Erwin has written a further ten books on puzzles and scientific subjects. In September 1995, he was awarded the 'Order of Merit in Gold' from the city of Vienna, in recognition of his literary achievements.

Contents

Introduction

Welcome to the past, present and future all rolled up into one brain-busting collection for your reading pleasure and puzzle-solving enjoyment. For the next 140-pages or so you will travel at the speed of light across the whole planet, dip in and out of the known dimensions and decend down the rabbit hole into the wibbly-wobbly worlds of space-time and parallel universes.

From magnetic forces to mystical paradoxes and star gazing to ghostly shadows, an entire spectrum of riddles about the very nature of our existence will be thrown into question. Are you ready to unlock their secrets?

On this adventure though time and space, you will encounter spaceships, super-computers and powerful mathematics, string theory, strange sounds and monster myths, all with dastardly-devised conundrums for you to solve, and most importantly, like a true scientist, show your methods and working.

You will come face to face with dark matter and invisible light, you will hold time bombs in the palms of your hands and you will wonder at the sheer complexity of our universe and this place that we call home. Within the very pages of this book, this world will become yours to peruse and ponder, a playground where your imagination can roam free. Before your very eyes, your brain will transform into an effective detective, searching for significant clues, hunting for answers, unable to rest until you find the clever culprits responsible. Are you up for the challenge?

Resist all temptations to turn to page 99, the Answers, no matter how tangled in knots, or baffled your brain becomes. Your task – should you choose to accept it – is to unlock the unique mysteries contained within each puzzle, one for every day, until you complete the entire book. Only then would you have truly mastered the art of problem solving.

Now, there's lots of puzzles to complete, so you better get cracking... GO!

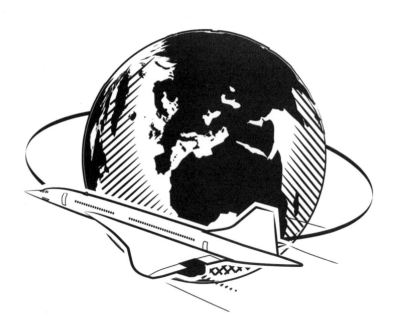

HOW DO YOU BUILD A TIME MACHINE?

Most readers will have heard of the International Date Line (IDL), established as an irregular line, drawn by convention through the Pacific Ocean, substantially along the 180th meridian. IDL marks the place where the date changes.

Crossing the line from west to east, travellers gain a day and, conversely, lose a day traveling east to west.

The position of the IDL has been arbitrarily designated and is partly curved to accommodate eastern Siberia, then bulges westward again in order to avoid crossing land.

Let us now assume that a super fast aircraft can take its passengers around the equator, circling the globe in six hours. Flying west to east it will cross the IDL four times in 24 hours and consequently lose four days. Starting, say, on March 18 it will land on March 14. Continuing the journey the passengers will go back into history, eventually reliving the birth of Jesus Christ. Traveling east to west they will travel into the future, giving them a decisive advantage in the lottery on one's return.

Have I managed to build a time machine, or is there something wrong with this reasoning?

Solution on page 101

SPACE INVADERS

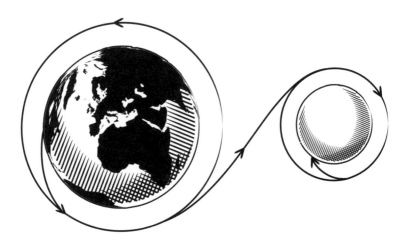

On July 20, 1969, Apollo 11 astronaut Neil Armstrong was the first human to descend a lunar ladder to the Moon's surface. Yet for a century or more a moon landing was a favourite topic of science fiction.

A German pre-war film *Die Frau im Mond* (*The Woman in the Moon*) was considered pure utopia, only 70 years ago. This takes us to an interesting problem.

When Apollo 11 was launched to the Moon, it travelled in a curve like the figure 8, as illustrated:

Why did Apollo 11 not go in a straight line?

Solution on page 101

THE FORCE

In many ways, magnetism and gravity are similar. For instance, they are both forces and they both obey the Inverse Square Law. Put simply this means: halving the distance between two objects quadruples the force between them.

Imagine you had a piece of iron and a very strong magnet. It is virtually impossible to put the iron on the magnet silently. At the last moment the magnet seems to snatch it away causing a click. Try putting a fridge magnet quietly onto the bare metal of your fridge. It's impossible.

However, it is very easy to put the same piece of iron on the ground without making a noise. Can you explain the difference?

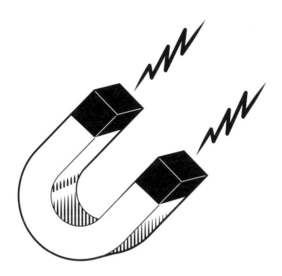

Solution on page 101

PLANET MYSTERY

The view taken by scientists, of our world since antiquity went, in stages, through a remarkable metamorphosis.

The Earth was first thought of as a disc in the centre of the universe. It was not until the sixteenth century that the spherical form generally accepted.

The Polish astronomer Nicolaus Copernicus (1473–1543) came to the conclusion that the suggestion mooted by some Greek philosophers, as early as the Third Century BC, that the Sun and not the Earth was the centre of our solar system, was correct.

This theory was proved by Galileo Galilei (1564–1642), whose prosecution and conviction for heresy, became one of the first examples of ecclesiastical miscarriages of justice.

For all of us who are not professional astronomers many intriguing mysteries remain to test our cognitive faculties.

Assume for the purpose of this mental exercise, that our globe, while remaining in orbit round the Sun, continues its current revolution on its own axis but once a year, instead of once a day.

The question is, what effect, if any, would this have on our physical world, with specific reference to:

(a) The length of our day
(b) The Moon
(c) Weight of objects
(d) Climate
(e) Satellites (geostationary orbit)
(f) Coriolis Effect

Solution on page 102

UFO ENIGMA

UFO stands for Unidentified Flying Object. Many thousands of sightings have been recorded during the last 50 years. Most of these can be explained by normal objects, aircraft, birds, cloud formations, but that still leaves a few hundred for which in-depth investigations by the U.S. Air Force can find no explanation. I presume that evidence will remain inconclusive until an extra-terrestrial creature actually descends the steps of their UFO and shakes hands with an earthling.

Until then, the first question arising is how such a craft would be propelled to account for the UFO's ability to hover, as well as move with fantastic speed.

H.G. Wells (1866–1946), the utopian writer, suggested one method – the gravity shield.

Let us assume that advanced UFO technology would indeed be able to reduce the Earth's gravity from an acceleration of 10 metres per second per second down to zero, covering the full range, so that the Moon's gravity comes into play. Would this solve the propulsion problem? In other words, could the UFO fly at zero Earth gravity, and using the Moon, move at great speed away from the Earth?

Solution on page 103

EQUINOX

At the equator the days and nights are always of the same length, 12 hours. At other latitudes the equinoxes (equal day/night lengths) occur twice a year, in the Spring and Autumn. This is because the Earth's axis is not perpendicular to a line joining the centres of the Sun and Earth. This causes the shadow (night) to be unevenly distributed between North and South; one hemisphere will have more day (Summer) and the other more night (Winter).

It is also a well-known fact that sunsets and sunrises happen more quickly at the equator.

Why?

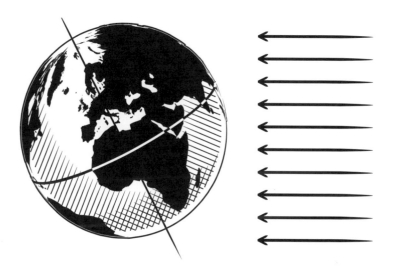

Solution on page 103

MOON BALLOON

The Moon has long been associated with mystery and romance, and it is not difficult to see why. It is a prominent object in the sky that shows changes in appearance so large and regular as to be noticed by even the most casual of observers. The connection with tides must have seemed magical to the primitive mind.

It is difficult to imagine a scene more emotive than a large full moon glimpsed through the dark silhouette of trees. Which brings me to my question: We now know that the moon is a rocky satellite, incapable of expanding and shrinking like a balloon, so why does it look larger when setting and rising?

Solution on page 104

SQUARE WORLD

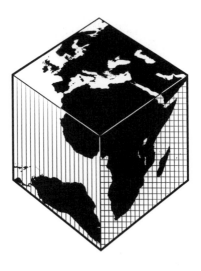

It might not seem like it to you and me, but our world is incredibly smooth. The vertical distance from the top of Mount Everest, Nepal, to the depths of the Marianas Trench, in the western Pacific Ocean, represents a distance that is only about one third of one per cent of the radius of the Earth. Mountains seem high to us, because we are relatively small.

It is lucky for us that the globe is not perfectly smooth, because the seas would cover the entire surface, and there would be no dry land.

I was wondering what the world would be like to live on if it was a perfect cube. Let us ignore the rotation for a moment. It would still have gravity, so it would be possible to walk around on each of its six flat surfaces. Imagine walking up to an edge and looking over. Would the second face seem like a gigantic cliff? And suppose someone on that face had walked towards the same edge. Would we be standing toe to toe?

Solution on page 105

STAR GAZING

I was walking home one dark evening with my two nephews, Harry and Tom. The night was very dark and the stars were very bright. Tom said, "Look, some of the stars are coloured. I can see blue ones and red ones".

"I know about this", said Harry, "It's called red shift and blue shift. What we call white light is really a combination of all colours of the rainbow. If stars are travelling towards us the light waves get compressed causing all the light to be shifted towards the blue end of the spectrum. And the opposite happens when the stars are moving away".

Tom thought about this for a little while before replying. "You must be wrong for two reasons. Firstly, the universe is supposed to be expanding, but most stars we see are white. Secondly, suppose there was a blue shift – all the light would be shifted. So violet would turn into invisible ultra-violet, indigo would become violet, and so on. The invisible infrared would turn into visible red. So, all the colours would still be there, and the star would still look white".

At this point they both turned to me. "Uncle...?"

Solution on page 105

BEFORE THE BEGINNING

In 1929, Edwin Hubble made the startling discovery that galaxies are all moving away from each other. Before this time, most people assumed that the Universe was a fairly static place; that although stars may come and go, the overall picture remained the same. Now all this changed.

Hermann Bondi, Thomas Gold and Fred Hoyle attempted to bring back some order by proposing, in 1948, their Steady State Theory in which new galaxies were continually being created to replace the receding ones.

The now generally accepted theory was proposed by the Russian physicist, Alexander Friedmann, who had predicted the expansion seven years before Hubble discovered it.

If one imagines the tape being played backwards, one could see the stars and galaxies converging to a point. This must have occurred between ten and twenty billion years ago. This was the Big Bang.

Two questions are very frequently asked by people who see the theory for the first time:

"What lay outside the point?"

and

"What was the Universe like before the Big Bang?"

Solution on page 106

A MASSIVE PROBLEM

Meteorites falling to Earth can give a spectacular light show. Very large ones could cause a great deal of damage, but are, fortunately, rare. However, the Earth is continually bombarded with a fine dust from outer space.

This must mean that the Earth is continually increasing in mass. Does this affect the Earth's orbit round the Sun?

Solution on page 106

POCKET MONEY

My friend Andrew is an incorrigible gambler. One evening, having lost at chess, he made me the following proposition:

"I hear some coins jingling in your pocket. I also have a number of pound coins. Let's count them and the one who has the larger amount must give them to the other. If there is a tie, the deal is off."

It so happened I lost, as I had ten pounds against his seven. It then occurred to us that this little bet presented an interesting paradox.

Each of us could argue as follows:

I know how much money I stand to lose, but on the other hand I stand to gain more than I have. Consequently the bet is clearly to my advantage.

Prove by unimpeachable logic why the argument that the bet is favourable to both gamblers is fallacious.

Solution on page 107

ELLIPSE

Prove by simple logic that no regular polygon of more than four sides can be inscribed in an ellipse.

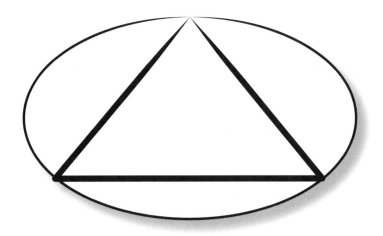

Solution on page 107

ONE SIDE OF THE MOON

It is common knowledge that we on Earth see only one side of the moon. In other words, the moon rotates at a rate around its axis, which synchronizes it with the Earth's rotation. Is this a coincidence?

Solution on page 108

KING OF CARDS

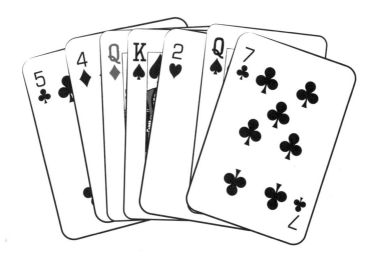

A bridge player selecting a number of cards from a deck makes the following statement: "All but three of my cards are Jacks, all but three Queens, all but three Kings, and all but three Aces."

What is the least number of cards he must hold to justify his statement, and what are they?

Solution on page 108

SUM OF ALL

If a and b are two different positive numbers, will $a^2 + b^2$ be larger or smaller than $2ab$? Prove your point.

12 99 12 86 06 31 1 21 23

16 100 42 5

49 42 164

6 1001 7 164

9 201 75 4 16

Solution on page 108

GRAVITY

An astronaut awakes in an enclosed room where gravity is apparently normal. He realizes that there are three possibilities:

1. He is subjected to gravity.
2. The spacecraft is accelerating.
3. The spacecraft is creating its own "artificial gravity" by spinning.

Is there any experiment that the astronaut can do inside the room to discover which is the true situation?

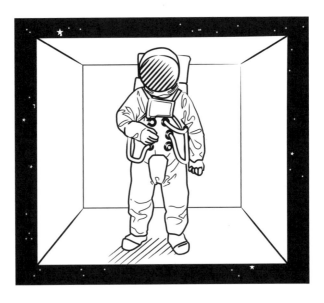

Solution on page 109

ZERO OPTION

Find the centre of gravity of the figure below. Here logic and not the Zero Option is the key.

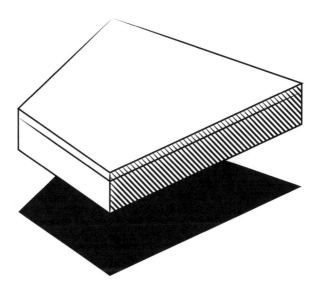

Solution on page 109

PULL OF
THE SUN

Surely the Sun's gravitational pull on the moon is much larger than the Earth's. Why then does the Sun not pull the moon away from its orbit around the Earth?

Solution on page 110

A GAME OF THREE SIDES

Prove that half the perimeter of any triangle (S ÷ 2) must always be longer than any of its sides.

Solution on page 110

CHANGING SIDES

You have a rectangle with sides a and b. If you increase a side by 1/3, prove that you have to decrease side b by 25 per cent in order to retain the same area.

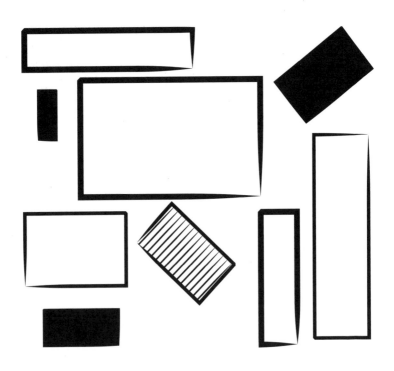

Solution on page 111

FOUR BY FOUR

Prove that the square, which is a special case of a rectangle, contains the largest area for a given perimeter of any quadrilateral.

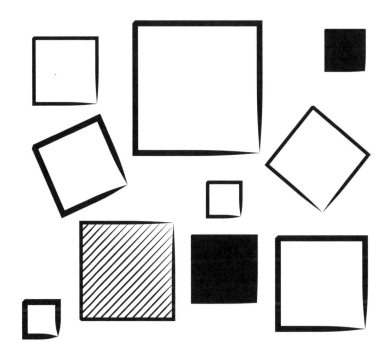

Solution on page 112

FRED & STEVE

Fred has two children. They are not both girls though one might be. What is the probability that both are boys?

Steve also has two children. The elder is a girl. What is the probability that both are girls?

Solution on page 112

ON THE TILES

A two-and-a-half inch square disk is thrown at random onto a tiled floor of a repeating pattern, as illustrated.

Prove that the odds against the disk definitely not landing across any line is at least one in 6.856 and explain why the odds cannot be precisely stated.

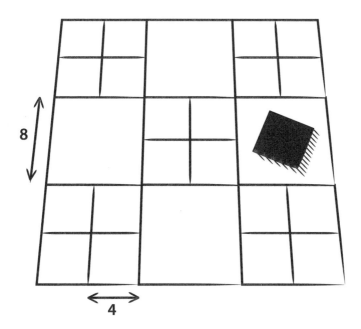

8

4

Solution on page 113

OLBERS' PARADOX

Heinrich Wilhelm Matthias Olbers (1758-1840), a German astronomer, hypothesized that if the universe were infinite and contained an infinite number of stars, then the whole sky should be as bright as sunlight. After all, the observable universe alone indicates that the Milky Way is only one of several hundred billion galaxies, with each galaxy containing in turn some hundred billion stars. It seems therefore inconceivable that any line of sight could miss a star.

You can compare it with a blackboard on which you are asked to make a mere one million chalk marks. Surely the blackboard would appear white? Why then do we not see the sky brightly lit at night?

Solution on page 114

BLACK TO THE FUTURE

Our modern world is full of light. Walk through any city centre at night and you will see all sorts of coloured lights, from white through to the full spectrum of a rainbow.

But it occurs to me that there is one colour that is omitted, and that is black. This seems a great pity, because I could think of many uses. For example, one could use a black bulb in a photographic darkroom, avoiding the necessity of costly blackout material. During a war, searchlights were used to illuminate bombers at night and also to dazzle the pilots. A black searchlight, however, could be used during the day to plunge the bombers into darkness.

What's wrong with this thinking?

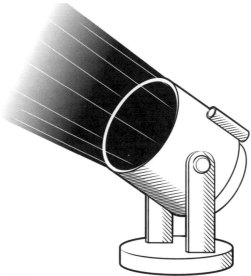

Solution on page 114

RINGING TONE

I was recently putting in some fencing in a field. While knocking in the fence posts, I noticed differences in the sound. In the centre of the field I was a long way from anything, and the sound from the hammering of the posts was dull and flat. I concluded that this was because there was nothing near to reflect the sound. In another part of the field I could hear a distinct echo from a nearby building.

However, in a third part of the field I could hear a distinct ringing tone. Can you think what might have been the cause of the sound?

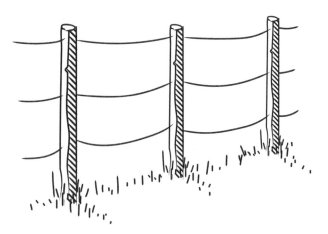

Solution on page 114

SOUND REASONING

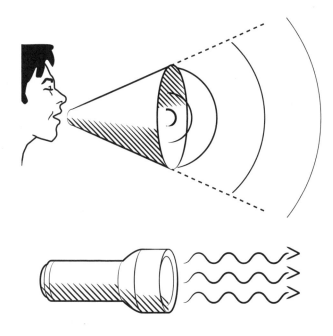

Most people know that light, like sound, travels in the form of a wave. Yet light and sound behave very differently. Can you explain:

1. Why it is you can hear around corners, but not see around corners?
2. Why you can use sound to cancel out other sounds, but cannot do the same with light?

Solution on page 115

FREQUENT CHANGING

On a recent drive around the country, I was listening to a very interesting programme on the car radio. As I moved away from a transmitter, the reception got progressively worse, so I had to retune the radio to pick up the same station on a nearer transmitter. I had to repeat this process several times during the journey. This really affected my enjoyment of the programme. Why do they not ensure that all transmitters broadcasting the same station use the same frequency?

Solution on page 115

LIGHT FANTASTIC

Imagine two blacked-out rooms bathed in yellow light. A piece of white paper examined in both rooms appears to be exactly the same shade of yellow. Another piece of paper appears black in one of the rooms and striped red and green in the other. How could this be?

Solution on page 116

FASTER THAN LIGHT

The other day, I was trying to explain relativity theory to a friend. Einstein had predicted, and experiments had confirmed, that as objects moved faster they also increased in mass. At the speed of light, any object would have infinite mass and so would need infinite energy to make it go faster.

"This means that there is a maximum speed at which anything can travel, and only massless things like light can travel at that speed," I explained.

My friend is very bright, and after a little thought he came up with three situations that seemed to contradict Einstein. Are any of them valid?

1. The poles outside barbershops have helical stripes painted round them. If the pole is rotated, its stripes are seen to travel along it. If the rotation is fast enough the stripes should travel faster than light.

2. If you shine a flashlight on a wall, you will see a spot of light. You can make this spot move along the wall by rotating the flashlight. The speed of the spot depends on the distance from the wall and the speed of rotation of the flashlight. If you shine a laser on a target some miles away, the spot of light formed could be made to move faster than the speed of light by rotating the laser.

3. Very small particles, such as neutrons, can penetrate matter very easily. Imagine such a particle, traveling at nearly the speed of light, entering a block of glass. The particle is so small that it will not be impeded by the glass, but light is slowed down to about two-thirds of its speed in air. Therefore, the particle would be traveling faster in the glass than the light.

Solution on page 116

LIFE IN TECHNICOLOUR

Rainbows, an enchanting spectacle of nature, pose a number of baffling questions. We know they are caused by drops of water falling through the air, refracting sunlight in such a way that it creates a curve of light exhibiting all colours of the spectrum in their natural order. Now answer the following:

1. Why is it you do not always see a rainbow when it rains while the sun shines?
2. Can the moon also produce a rainbow?
3. Even without rain you can, at times, see a rainbow if you look across a lawn early in the morning. Why?

Solution on page 116

COLD FEET

The other night I could not sleep and was feeling hungry. Deciding to raid the refrigerator, I stepped, barefoot, from the carpeted hall into the tiled kitchen and found that the floor was extremely cold. The door between the hall and the kitchen had been open all the time. So why was the kitchen floor so much colder than the hall floor?

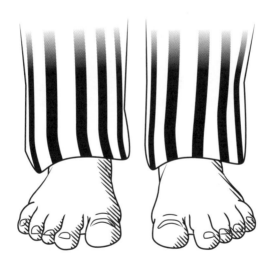

Solution on page 117

CAR HEADLIGHTS

Many driving accidents at night are caused by the glare of headlights from approaching cars. My friend Jonathan, who fancies himself as an inventor, came up with an idea offering an effective and inexpensive solution to the problem.

"Let us introduce polarizing filters in front of headlights," he said, "to polarize light horizontally, which would absorb all photons whose electric vectors are vertical. Conversely, let us use windshields with a filter turned 90° to the first, absorbing light emitted by the headlights."

On the face of it, this would be the perfect solution, as the light from the approaching car would be blocked out, while all other objects would be visible.

Would this idea work? If so, why hasn't it been adopted?

Solution on page 117

VISION
AND SOUND

Why are humans endowed with two eyes and two ears? Is the second organ just backup equipment or a spare, or has it another function?

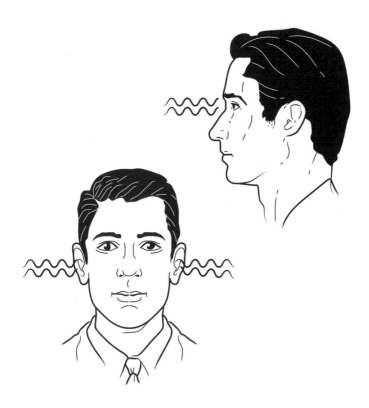

Solution on page 117

TOY BOAT

As a boy I built my own little steamboat to operate in the bathtub. It was my own design, using odd bits of metal plus a little fatherly help, and looked something like this:

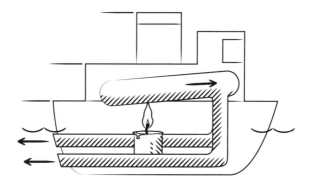

The principle was simple. The candle heated the water in the boiler and the resulting steam forced the water through the outlets as jets, propelling the boat forward. To my surprise, I noticed that, from time to time, the outward jets stopped and instead water was sucked in through the tubes. I did not understand why this should be and, more surprising, why the boat did not then reverse its direction and move backwards, as a basic law of physics would suggest (Newton's Third Law of Motion: action = reaction).

Can you explain?

Solution on page 118

BREATH ON YOUR HAND

If you breathe gently on the back of your hand it feels warm.

If you purse your lips and blow hard on the back of your hand it feels cold.

Surely the temperature of one's breath is the same on both occasions, so why does it feel different?

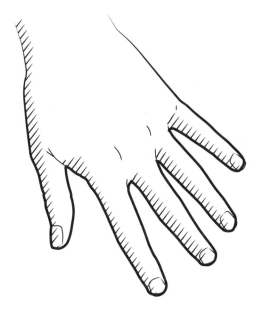

Solution on page 118

THE FOEHN

It has long been accepted that weather has a profound effect on the physical and emotional well-being of most people. One of the well-known natural phenomena is a warm dry wind blowing down from high mountains into valleys. In German-speaking countries this wind is called *foehn*, though it has many other names with one thing in common: dry, warm and unpleasant. *Foehn*, or its equivalent, is held responsible for severe headaches and, in the extreme, for criminal behaviour in some of those affected by it.

How can a warm wind come down from a cold mountain, reach speeds of close to 75 mph, and have such a physiological impact on many people?

Solution on page 119

DRINKING BIRD

Many toys and promotional gadgets are based on physical phenomena, often not readily understood. The drinking bird is one of the most fascinating examples. It consists of a glass bird, standing in front of a container filled with water.

The bird appears to be thirsty at regular intervals and so it rocks forward and dips its head into the water. Having quenched its thirst, the bird rights itself, only to repeat the performance after a little while. All you have to do to get the operation going is to wet the felt that is covering the bird's head and beak, after which the bird continues to bob up and down without further assistance.

What makes it do this?

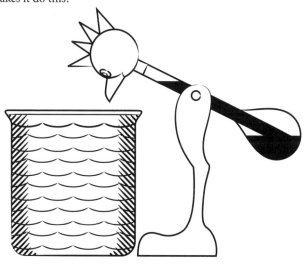

Solution on page 119

TV INTERFERENCE

A friend of mine lives east of Heathrow, in the middle of the approach path of planes coming in to land at the world's busiest airport. He installed double-glazing windows to reduce the noise of the planes, but finally lost his patience with the irritating interference and static on his TV picture.

Why should a passing plane cause this interference and how does it manifest itself?

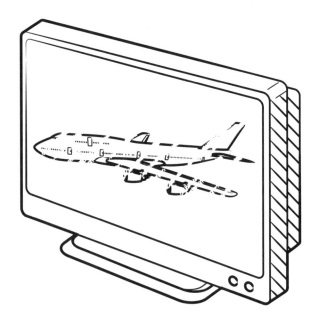

Solution on page 120

INVISIBLE MAN

In 1933 a movie with title *The Invisible Man*, with Claude Rains, was released. It is an age-old dream to invent a magic hat that will turn us invisible at will. However, to avoid the problems confronting Claude Rains in the movie, the process must be reversible. Our imagination runs wild, thinking of the possibilities that such a "fly on the wall" invention would offer. We know that sound can be extinguished by destructive interference, which occurs when two intersecting waves of the same frequency are completely out of phase, i.e. when the crest of one wave coincides with the trough of the other, in which case the two waves cancel each other out.

Can the same principle be used to generate destructive light interference thus rendering objects invisible?

Solution on page 120

RED EYE

I often wonder why the photographs that I take never look as good as those taken by a professional photographer. Even if I use an expensive camera, the results seem quite disappointing.

One thing that I cannot be blamed for is the "Red-Eye" effect. This is the ghastly appearance of people who are looking at the camera when I use a flash. Their eyes glow spookily red. What causes this effect, and how can I avoid it?

And while you're pondering this question, here's another one…

It is said that the reason why human eyes glow red is because of the rich blood supply at the back of the eye. But this does not explain why cats' eyes glow green.

Why do you think this is?

Solution on page 121

LOOKING BLANK

The colours in a television's picture are produced by combining the various brightness of the primary colours red, green and blue. For instance, red and green dots viewed from a distance appear yellow. If all the pixels are on full brightness, that part of the screen appears white.

If you were asked what colour a TV screen is when the set is switched off, you would probably say that it is grey. So, here is my question: how does a TV display black? Presumably turning all the colours in the required area off produces this 'colour', but this should produce grey.

Solution on page 121

ENLIGHTENMENT

Candle flames have long held a romantic and scientific fascination. We have all stared into the guttering flame, and few can resist the temptation of playing with the molten wax.

Here is a deceptively simple question: The candlewick is burning, and so the hottest part of the flame is, presumably, in the centre. However, candle flames are darker in the centre and lighter on the outside.

Why is this?

Solution on page 122

GHOSTING

I am familiar with the effect called "ghosting", which can give one a poor television picture. It is caused by a nearby building or structure reflecting the TV signals. This means that the picture is received twice, once from the main beam, and then, a short while later, the second weaker signal arrives. This causes a ghostly second picture to the right of the main one, the distance between the two pictures being determined by the position of the reflector.

The other day I happened to glance out of a window, and I was astonished to see a very similar effect directly. Across the road a man was enjoying a walk in bright sunshine, behind him was a dark wall. A few feet in front of him I could see a ghostly copy also walking along.

Can you explain how this happened?

Solution on page 122

TRAVELLING LIGHT

Galileo Galilei is well known for his work in astronomy. It is less well known that he also devised an experiment to measure the speed of light. This was based on an earlier measurement of the speed of sound using distant cannons. He had an assistant stand on a nearby mountain at night with a darkened lantern. Galileo then flashed his light at his helper who immediately signalled back with his lantern and measured the time it took the light to travel the distance and back – about three-quarters of a second.

If the distance had been doubled, do you think that the time would then have been:

a) three-quarters of a second
b) one and a half seconds
c) three seconds
d) or something else?

Solution on page 123

LOOKING GLASS

For centuries people have been fascinated by the mysterious world that seems to exist on the other side of the mirror. So like our own, yet made subtly different by the inversion of left and right. The glass seems to be an impenetrable barrier that prevents joining our looking glass world. Physics tells us that the image of an object appears to be as far behind a mirror as the object is in front. Yet I know that it is possible for the image to be in front of the glass, and to exist in our world along with the real objects.

How can this be so?

Solution on page 123

FLYING IN A CIRCLE

It was one of those hot, sleepy afternoons. I was stretched out on a sofa and, glancing in a mirror, I happened to see the image of a bird flying by. It occurred to me that I would have seen the same effect if the bird had been stationary, and someone had rotated the mirror. I then began to wonder what the effect would be with the bird and the mirror both moving. Could one effect cancel out the other? I then considered the following "thought experiment", which ruined my afternoon's rest.

Imagine that a bird is flying in a perfect circle round a mirror at the centre of that circle.

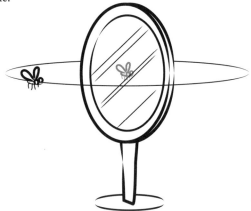

It should be possible to rotate the mirror at such a speed that the image of the bird appears stationary. What would be the relationship between the speeds of rotation of the mirror and the bird?

Solution on page 124

THE WINE GLASS

A well-known parlour trick is to slide a finger around the edge of a wine glass. After a few seconds, and a bit of practice, the glass starts to emit a weird sound. The finger has to be moving at the correct speed. If you move your finger faster, you might have expected the note's pitch to change but the glass just stops singing. What causes this sound?

Why is the pitch of the note always the same? And why does your finger have to be wet?

Solution on page 125

TAKE A BOW

"If music be the food of love," Shakespeare commands, "play on."

My favourite group of instruments are chordophones, or stringed instruments. Some are plucked, for instance, a lute, guitar or harp and some are bowed, such as a violin or viola.

What intrigues me is this:

The fact that plucking a string produces music is not surprising, as the source is clearly the vibration of the string. But how does bowing an instrument produce sound? Surely the smooth action of the bow cannot excite the strings to vibrate.

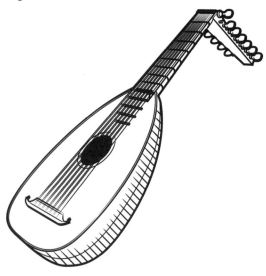

Solution on page 125

STRING THEORY

The origin of the word 'orchestra' is somewhat incongruous. It comes from the Greek word for a place intended for dancing.

An orchestra consists of a large array of various types of instruments, including strings and winds performing in sync. The remarkable phenomenon is that the pitch of most instruments change as the orchestra tunes up.

Can you describe the changes in the instruments listed and explain the reasons?

Solution on page 126

SNOW LANDSCAPE

It was snowing heavily all night. When I left my home, the traffic was as heavy as on any other weekday and yet the usual noise was muted as if someone had turned the volume control slightly down.

Perhaps the snow had something to do with it, but if so, why?

Solution on page 126

FREQUENCIES

Have you noticed how strange your voice sounds when you play back your own recording. And yet, when you listen to other people's recording, their voices sound quite normal.

Why should only your voice be distorted?

Solution on page 126

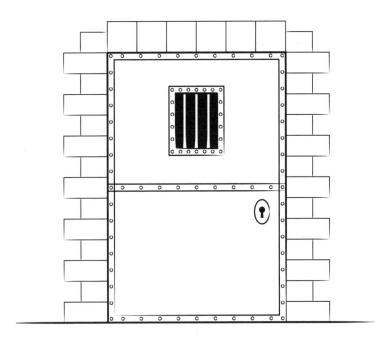

DUNGEONS OF DIONYSIUS

Dionysius The Elder (432-367 BC), the tyrant of Syracuse, is known to have constructed his underground prisons in accordance with certain acoustic principles, which enabled him to listen to the conversations of his prisoners.

Is this merely a myth or are such constructions possible?

Solution on page 126

SOUP ON THE ROCKS

I am too impatient to wait until my scalding hot soup cools to a manageable temperature. To speed it up I drop an ice cube into the soup. I would have expected the cube to melt slowly and noiselessly, but to my surprise I hear a sound like the cracking of a whip.

What causes this intriguing sound?

Solution on page 127

THE
BLOWING HORN

When a car, with its horn blaring loudly, passes you at high speed you will notice a distinct shift in pitch, increasing on approach and reducing as the vehicle zooms passed.

This is perhaps surprising as the speed of sound, and its frequency is a constant for any medium.

So, why does the pitch of the noise change as the car travels passed you?

Solution on page 127

WALLS HAVE EARS

Nothing is more annoying than noisy neighbours. It seems that no insulating material is effective at deadening the noise from a deafening stereo system. But I have an idea.

I know that microphones convert sound energy into electrical energy. Therefore, if I embedded thousands of them in the walls, floors and ceilings, they will convert the aggravating and useless sound into electrical energy. I might be able to run something useful with it like my kettle or a light bulb.

Would my idea work?

Solution on page 129

SPEED OF SOUND

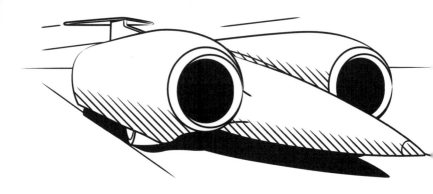

In October 1997 the jet powered car, ThrustSSC, piloted by Andy Green, became the first land vehicle to travel faster than the speed of sound, 760mph.

This is a remarkable achievement. But can you think how a car travelling at half that speed, a mere 380mph, could also claim to have travelled faster than sound without actually breaking the sound barrier?

Solution on page 128

TIDAL POWER

Africa is a very large continent. It is almost 5,000 miles from Senegal to Somalia. This means that the state of the tides on the east coast will be very different from that of the west coast. Imagine a large pipe is constructed to connect the Atlantic and Indian Oceans, with a power station built in the middle, and water flows from the high to low tide areas in order to drive turbines and produce electricity.

Would such a scheme work?

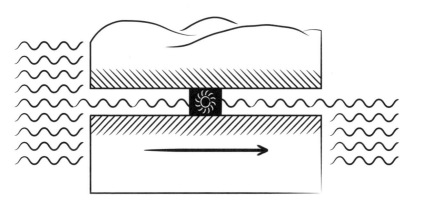

Solution on page 129

SUPER COMPUTING

I was once walking through the city of Dublin when I saw a fortune-teller sitting by the side of the road. He had a sign with his scale of charges: £2 for a palm reading, and £5 for a Tarot Card reading. I wondered why there was a difference in price. Was one method supposed to be more accurate than the other? Was one more difficult, in which case why bother with it? The man did not look as though he would invite a philosophical discussion, and I foresaw an argument ensuing if I asked, so I walked on.

The incident set me thinking about how knowledge is associated with power. Accurate prediction of future events has always been and will continue to be an extremely valuable asset. Modern computers are excellent for rapidly performing complex calculations, such as predicting weather for the following day. Of course, the more accurate the data, the (more) better the prediction. I wonder if there is any limit to this.

Computers are becoming increasingly powerful. Let us imagine that I had access to one of the super computers and I fed the machine all the information about a particular horse race; everything about each horse and rider, each blade of grass on the racetrack, the weather, the spectators – everything. Could the computer accurately predict the result of the race?

Solution on page 129

CITY SPEED
LIMIT

George Gamow's wonderful book, Mr Tomkins in Paperback, is about a bank clerk attending a number of scientific lectures. He finds it difficult to concentrate on the lectures and soon drifts into an alternative dream world.

During one such dream, Mr Tomkins found himself in a city where light travelled at just a hundred miles an hour, instead of the usual 186,000 miles a second. At first sight everything looked normal, but he soon encountered some strange phenomena. For instance, at the railway station a strapping young man who was just about to board a train was wishing farewell to a very old lady. "Goodbye, Grandfather", the ancient woman said.

What was the subject of the lecture, and how could the grandfather be younger than his granddaughter?

Solution on page 130

DARK MATTER

We are surrounded by solid objects. But appearances can be deceptive. We know from the study of atoms and molecules that they are mainly empty space. The nucleus of an atom occupies about one hundred thousandth of the diameter of an atom. If an atomic nucleus were scaled up to be the size of a golf ball, the nearest orbiting electron would be about one mile away. If all the empty space were squeezed out of the world, it would be about the size of an orange. So, although things may look solid, they could be considered as ghostly objects, not really there. This is my question:

If I am mostly empty space, and the wall next to me is also hardly there, why can I not walk through walls?

Solution on page 130

REVOLUTIONS

There is no easy way to store large amounts of electricity. This means that the amount of electrical energy generated by the power stations is always equal the energy consumed plus any heat loss during transmission.

Imagine an anarchist group in a country with an unpopular government planning to use this as a weapon against the State. All the members of the group, let's say a million of them, could turn on all their electrical devices, causing all the power generators to work hard to satisfy the demand. Then at some predetermined moment, everyone would switch everything off, simultaneously.

Because of the inertia of the system, the power would continue to be produced but not consumed. The power stations would all blow up, dealing a devastating blow to the Government.

Would this dastardly plot work?

Solution on page 131

TURNING ON THE POWER

There is a great interest at the moment for renewable, pollution free sources of power, and I would like to add my suggestion for consideration.

A gyroscope is a heavy flywheel that rotates at high speed. Such a device stores kinetic energy and is resistant to dislocation because of inertia. They are used in aircraft as compasses. The gyroscope is mounted on gimbals, and whichever way the plane heads, the gyroscope always points in the same direction.

Imagine an enormous gyroscope set up on the North Pole parallel to the ground. It would try to continue to point in the same direction while the Earth rotated under it. So, if it were connected to the ground via a generator, the gyroscope would turn the dynamo generating electricity.

Is my proposition theoretically sound?

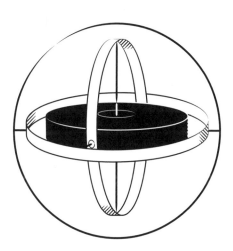

Solution on page 131

CLOCKING OFF

It always amazes me how many devices these days use digital clocks and timers – computers, central heating systems, cars, ovens, smartphones, and so on. All of them need to be adjusted twice a year when the clocks are put on an hour or back an hour. One of them always manages to elude me, and then I find I have caused some domestic disaster, like getting up an hour later than planned.

We are told that this process of Daylight Saving Time is to wisely use the daylight that is available to us. I would like to make the suggestion that we return to the system employed by the ancient Egyptians. They used to define sunrise as 6am and sunset as 6pm, the time in between being divided into twelve equal parts.

Would this system have any unexpected consequences?

Solution on page 131

THE TRANSPORTER

I love visiting different places, but hate the process of travelling; packing cases, catching planes and trains at specific times, keeping track of my luggage, and unpacking. I love the idea of a transporter that would instantaneously get me to where I wanted to go, bsut I worry about the practicalities.

In science fiction films such as Star Trek, the theory appears to be that the traveller is disassembled molecule by molecule and their body is transmitted to a destination thousands of miles away, where the individual is reassembled, again, molecule by molecule.

I could accept that the arriving individual would think that he was the same person that set off, because he had all the same memories, but is it not the reality that we killed the first person and made an exact clone, who was under the impression that he had travelled? Would there be any way in which we could tell if it was the same person or not? And if we could not tell, would it matter?

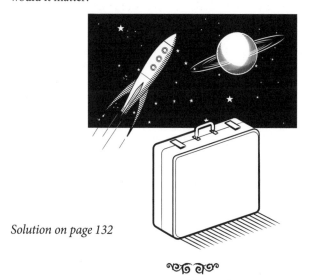

Solution on page 132

ULTIMATE
WEAPON

David Stone, in a letter to *Geotimes* (1969), describes a geophysical weapon, which if used by a nation with a large population, could have a devastating effect on an enemy.

Suppose, for instance, the residents of China, who now number over one billion people, were to all jump in unison on pogo sticks. Such an action could set up shock waves that could amplify with each jump, reaching perhaps a magnitude of five on the Richter Scale. This could, for example, destroy parts of the United States. To protect itself, the USA might organize jumps timed to cancel the offensive waves. But, considering the vast difference in populations, the Americans would either need to jump from a greater height or call on the NATO allies for jump assistance.

Before we get ahead of ourselves, would this Chinese super-weapon actually work?

Solution on page 132

CHARGING FOR A FLIGHT

I know from my school physics classes that whenever a conductor moves through a magnetic field, a voltage is produced across it. This was discovered by Michael Faraday in 1831 and is the basis for all electric generators.

It occurs to me that most aircraft are made of metal, and they move through the Earth's magnetic field. I did a calculation to see how much electricity could be generated. Unfortunately, the Earth's magnetic field is very weak, so even an Airbus A380 – the world's quickest passenger plane – would generate only about a quarter of one volt between its wingtips. But this might be enough to run some small electronic devices without the need for batteries or generators.

Is this scheme feasible?

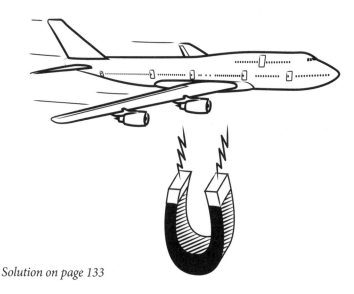

Solution on page 133

THE SCIENTIST

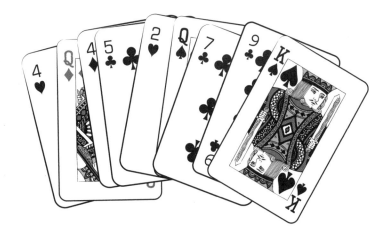

To test his logic and wits, The Scientist has a new card puzzle for The Gambler. The Scientist picks four cards out of the pack of 52 cards and lays them face down on the table. He offers four hints to The Gambler:

1) The left card can't be greater than the one on right.
2) The difference between the first card and third card equals to eight.
3) None of the Ace is present
4) No face card has been included, i.e. no Queen, King and Jacks.
5) The difference between the second and fourth card is 7.

The Gambler smiles and tells all the four cards to The Scientist.

What are the four cards?

Solution on page 133

TIME BOMB

Here is an idea for a new type of bomb that requires no explosives.

Imagine that the bomb was made of a material that was a perfect reflector of sound. Inside it has a spherical cavity. At the centre of this cavity sits a ticking watch. The sound inside will build up, as it cannot escape. Eventually the sound energy would be so great that it could no longer be contained. At this point the bomb would explode.

Would this design work?

Solution on page 134

THE RALLY

In training for the Monte Carlo Rally, Andrew drove from Basle to Lugano, a distance of 305km. He set out in his Jaguar XJ6 4.2 at 8am one day and arrived at his destination at 3pm the same day, clocking an average speed of 61km/hour; not bad, considering that he had to climb the St Gotthard Pass, with its numerous hairpin bends.

After a good night's rest, he started on his return trip at 8am, taking it very easy. He stopped off in Brunnen for a leisurely lunch and then again in Liestal for tea, arriving in Basle at 4.30pm.

Will there be a spot between Basle and Lugano where Andrew will have been at precisely the same time of day on both journeys?

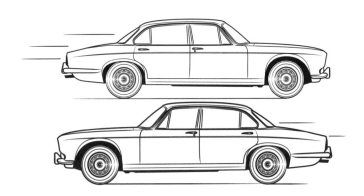

Solution on page 134

THREE CITIES

If City A is 9,000 miles from London and London is 9,000 miles from City B, prove by reasoning that City A must be closer to City B than 9,000 miles.

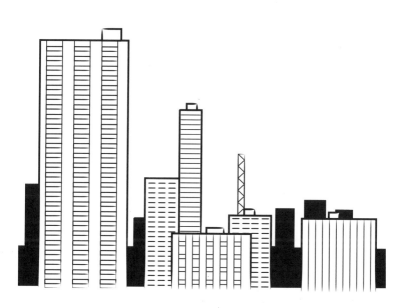

Solution on page 135

PIGGY BANK

Ever since my first birthday, my father fed a piggy bank with one pound for each year of my birthday: £2 on my second birthday, £3 on my third, and so on. Yesterday was another one of my birthdays and, as I needed some extra cash for a vacation abroad, I emptied the piggy bank and found that I had accumulated £300.

Can you tell me how old I was yesterday, and prove with a simple formula that you can find the answer, without going through the wearisome process, unworthy of an intelligent puzzle-solver, of adding 1 + 2 + 3 + 4... + N, until you reach 300?

Solution on page 135

THE PRICE
IS RIGHT

Most readers will have a basic understanding of the equilibrium price at which demand and supply of a commodity meet, determining its price.

Let us look, for example, at a typical demand and supply schedule for bananas:

Price	Market Demand per Pound	Quantity Supplied
10	800,000	0
20	180,000	0
30	160,000	70,000
50	100,000	80,000
70	90,000	90,000
90	60,000	120,000

It is shown that at a price of 70, demand and supply are in balance, establishing the price for bananas at that particular time. This can also be demonstrated graphically:

Your task. Prove and demonstrate how demand and supply curves can be drawn for the auctioning of the Mona Lisa by Leonardo da Vinci.

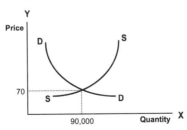

Solution on page 136

POND LIFE

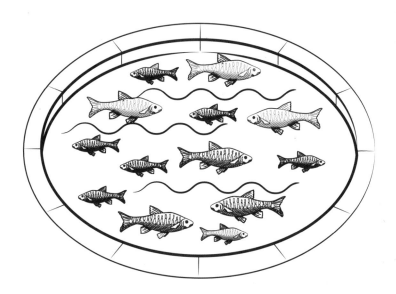

Your neighbour has a circular pond in his garden teeming with fish, impossible to count. Yet he tells you one day that he has 1,800 of them, give or take a hundred or so.

How can you prove him right or wrong?

Solution on page 137

COMBINATION LOCK

As part of an IQ test, I was given the following problem.

My prospective employer, David, pointed to the safe in his office, which had a combination lock with five digits 1 to 5 inclusive. My task was to find the correct combination in as few guesses as possible.

I ventured "1, 2, 3, 4, 5" on the assumption that one would chose a combination easily remembered.

"Wrong," David said. "In fact, not one of the digits is in the right position, nor is any digit suggested by you, following the one ranking behind it."

I considered whether this clue was sufficient to provide the answer. It escaped me. I then tried my luck with "4, 1 5, 3 2".

"That is better," David said. "You now have two digits in the correct positions and you have also two following the digit ranking before it."

Prove that these clues will lead to the only possible solution, which you are asked to find.

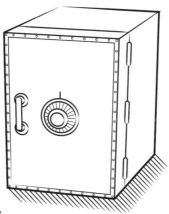

Solution on page 137

IMITATION OF LIFE

An art dealer has a number of paintings and is serving five customers. To the first he sells half his stock plus one painting. He continues the same formula until again he sells to the fifth customer half his remaining stock plus one painting, whereupon he is sold out.

How many paintings has he started with, and can you construct a formula that will give you an immediate solution for any number of customers?

Solution on page 138

THE MANUSCRIPT

Harry Spicer, a best-selling author, was given a deadline by his publisher to finish the last 80,000 words of his latest novel. "No problem," Harry thought, "if I write 4,000 words a day I shall be all right."

However, with one distraction or another, his output for the first half of the manuscript was only at the rate of 2,000 words a day. "No matter," Harry said to himself, "if I write 6,000 words daily for the second half of the manuscript my average will still be 4,000 words a day."

Was he right?

Solution on page 138

CENTROID OF A TRIANGLE

If you consider a triangle to be not just a geometrical concept but a physical object made of homogeneous material of uniform thickness, for instance of paper, it will have a centre of gravity (centroid) that can be ascertained with compasses and an unmarked straightedge.

Draw triangle ABC and join the vertices to the midpoint of the opposite side.

Prove that those lines will concur at M, and that M is the centre of gravity of triangle ABC.

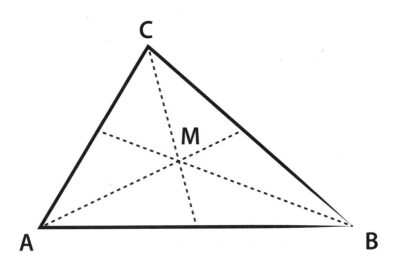

Solution on page 139

THE UNWILLING TRIANGLE

There are a number of right-angled triangles whose sides are whole numbers as can be shown by Pythagoras.

Examples: $3^2 : 4^2 : 5^2$

$5^2 : 12^2 : 13^2$

Can you prove that no isosceles triangle, such as the one shown below, can have whole numbers for sides?

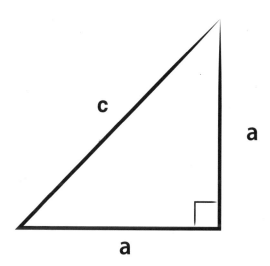

Solution on page 139

FIBONACCI'S SEQUENCE

Leonardo Fibonacci (1170-1250), also known as Leonardo of Pisa, was one of the outstanding mathematicians of his period. He contributed to recreational mathematics and discovered the Fibonacci Sequence, which is a series of digits where each term is the sum of the two preceding terms. Thus:

$$1, 1, 2, 3, 5, 8,$$
$$13, 21, 34, 55,$$
$$89, 144, 233 \ldots$$

You will find that the sum of the first 10 numbers, 143, is eleven times the seventh term.

Your task: Prove that this holds good for any Fibonacci sequence starting with any two consecutive Fibonacci numbers selected at random.

Solution on page 140

IT MUST BE EVEN

The other evening I was playing with the exponential button on my calculator and raising numbers to powers:

e.g. $3^7 = 3 \times 3 \times 3 \times 3 \times 3 \times 3 \times 3 = 2187$

and $3^5 = 3 \times 3 \times 3 \times 3 \times 3 = 243$

I noticed that if you add or subtract two powers of a number the result always seems to be even.

e.g. $2187 + 243 = 2430$ (even)

and $2187 - 243 = 1944$ (even)

Prove that whatever number you start with, the sum of any two of its powers must always be even, and that the difference between any two powers of the number must also always be even.

i.e. $x^a - x^b = 2n$ (where n is an integer)

$x^c + x^d = 2m$ (where m is an integer)

Solution on page 140

DISCOVERY

Fred and Harry are keen problem solvers, sharing an ambition not only to solve but also to construct challenging puzzles.

Playing around with numbers, they discovered an interesting phenomenon, namely that $5^2 - 4^2$ yields the same result as the sum of their bases. Is this a coincidence?

They try again with $13^2 - 12^2 = 25$. Indeed, $13 + 12$ is also 25.

Can you prove that this applies generally,

i.e. $a^2 - b^2 = a + b$, as long as $a = b + 1$?

Solution on page 141

PERFECT TIME

The Hour Hand (HH) of a lady's wristwatch is ¾cm, and the Minute Hand (MH) is 1cm long.

Prove that the endpoint of the MH moves 16 times faster than the one of the HH.

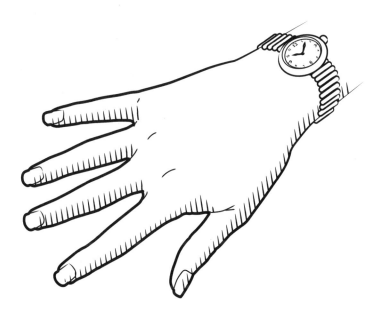

Solution on page 141

UNUSUAL SPHERES

I was wandering through an exhibition of spheres of all sizes and colours when I came across a sphere made of crystal glass, on a pedestal lit by a crimson spotlight from above.

A beautiful sight, to be sure, I thought. "But what is so special about this exhibit?" I asked the attendant.

"Well," he said, with pride in his voice, "this is the only sphere we have with the surface area and the volume the same four-digit integers multiplied by Pi."

"Remarkable," I muttered, trying to estimate the radius of this unusual object.

Can you find the sphere's radius?

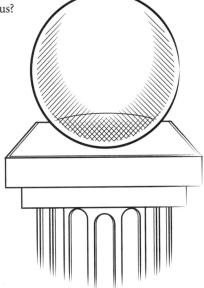

Solution on page 142

THE BILLIARD TABLE

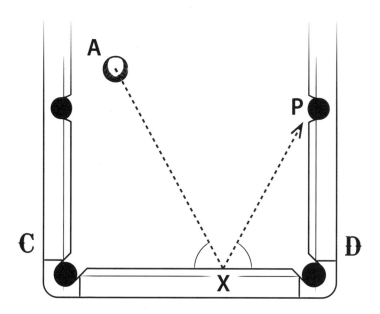

Starting at point A you want to project the ball into pocket P.

We know from our schooldays that the angle of incidence equals the angle of reflection. Accordingly, the ball will hit the pocket if angle AXC = angle PXD.

However, an expert player can, by spinning the ball, get it to rebound at a different angle and still make it land at P.

Prove that the original path from A to P bouncing off CD at X represents the shortest distance.

Solution on page 142

BULLSEYES

A target at a rifle range has a bullseye of 6-inches diameter and the next two rings with diameters 12-inches and 18-inches respectively. The bullseye scores 90 points, the 12-inch circle and the 18-inch circle score 42 and 27 in that order.

Within 10 shots, work out the probability that you will score 390, assuming that you will always hit the target at random. Ignore the fact that in real life such randomness does not exist – the shooter will always aim at the centre.

Solution on page 143

ANSWERS

How Do You Build A Time Machine

Yes, there certainly is. First, let us distinguish between natural time and an arbitrary time zone. A time zone is an area within which everyone has agreed to use the same time. However, every point on the Earth's surface has its own natural time. As one travels east, the natural time is farther ahead, 1 hour for every 15 degrees of longitude.

Let us suppose that the aircraft travels eastwards around the world so fast that it would complete the journey in one second, and it started from a point just to the east of the International Date Line. As it travelled it would encounter a natural time, which was progressively further ahead.

When it arrived at a point just to the west of the International Date Line the time would be 24 hours ahead. This extra day is then lost as the date line is crossed. The final time is then exactly one second after the start.

One cannot, therefore, travel into the past or future by using the International Date Line.

Space Invaders

When an object is thrown, it follows a natural curve called a trajectory. For it to deviate from this path, energy would have to be expended applying a force to it. The Apollo 11 spacecraft was, in effect, thrown by applying an initial thrust and then allowed to drift through space with only occasional corrections. The figure of eight curve is the natural trajectory in this case.

The Force

The Inverse Square Law predicts that as the piece of iron approaches the magnet, the force rises exponentially. In the last few millimetres, the force increases rapidly, making it very difficult to control.

Gravity also obeys the same law, but what is important here is the distance between the centre of mass of the Earth and the object. The Earth's centre of mass is a long way from the surface. Moving the object from, say, a metre, down to the surface means that the separation of the Earth and the object has changed by a mere 1 part in 6 million, causing a change in weight of only 1.000 000 3. So for all practical purposes, the force of gravity has not changed, making it easy to manipulate the object.

Planet Mystery

a) If one side of the Earth always pointed towards the Sun, this would be very similar to the situation with the Moon where one side always faces the Earth. One hemisphere would therefore be in perpetual day and the other everlasting night.

b) The Moon's orbit would be unaffected.

c) At the present, objects weigh slightly less at the equator than at the poles for two reasons. Firstly the Earth is slightly flattened at the poles (because of the Earth's rotation) and therefore objects at the Equator are slightly further away from the centre of the Earth. Secondly, the rotation of the Earth makes objects weigh slightly less, anyway. If the Earth took one year to rotate, both these effects would be much reduced, so objects would weigh the same at all points of the Earth's surface.

d) The climate would be drastically different. The dark side would be a few degrees above Absolute Zero, while the other side would become unbearably hot. Life would be impossible in these conditions. In theory there may be a ring in perpetual twilight where life would be possible, but it is unlikely that such a small area could support a full ecosystem.

e) Geostationary orbits are when the period of rotation of the satellite is the same as the time it takes for the Earth to spin once. For this to occur with a period of one year, the satellite would have to orbit the Earth at a distance equal to that of the Sun. This may be possible for a point on the dark side, but the Sun's own gravity would interfere too much on the light side.

f) The Coriolis Effect is caused by the rotation of the Earth. If the world were spinning very slowly, this would be very much reduced.

UFO Enigma

It could not. The resulting acceleration due to the Moon's gravity, which to start with is only about one sixth of Earth's gravity would only be about .03mm per second per second.

For the reader who wants to see the proof, here it is:

Gravitational force $= GmM$ (Newton's Law of Gravitation)

$r^2 = (6.673 \times 10^{-11} \times m \times 7.35 \times 10^{22})/(3.844 \times 10^{8})^2$
$= 3.3 \times 10^{-5}$ m
$= ma$ (Newton's Law of Motion)

therefore $a = 3.3 \times 10^{-5}$ m/s/s
$= .03$ mm/s/s

Equinox

At the Equator the Sun always sets (and rises) at an angle that is close to being a right angle to the horizon, at other latitudes the sun moves obliquely. When setting or rising vertically, it takes the minimum length of time for the Sun's complete disc to move from just touching the horizon (1) to just setting (2). After a sunset, the sky is still light for a while as the sun is just below the horizon (3). At northern and southern extremes of latitude, the Sun stays just below the horizon for an appreciably longer time.

In the illustration, the Sun takes three units of time to set to complete darkness at the Equator and about five at the other latitude.

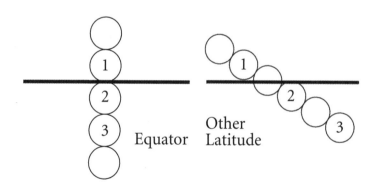

Moon Balloon

This is not a physical phenomenon. Measure the size of the Moon overhead, and near the horizon, and you will find they are the same. The effect is caused by our perception.

Most people know that when viewed from Earth, the Moon and Sun are apparently the same size, although few people could accurately estimate their subtended size. Try this for yourself: ask somebody to tell you, without trying it, which of the following objects, when held at arms length would appear to be about the same size as the Moon:

a) a soccer ball,
b) an orange, or
c) a drawing pin

Very few people would correctly select the drawing pin. We might like to think of our vision of the world as being absolute, but it is not. Our brain is continually interpreting what we see. The image at the back of our eye of someone walking towards us is continually enlarging as they get closer; but we do not see them 'growing' – we see them 'approaching'. The brain makes decisions using memory and cues of nearby objects. When this goes wrong we call it an 'optical illusion'. When the Moon is overhead, there are no nearby objects to compare it with, whilst near the horizon there are. Our brain perceives it to be much closer than it actually is; like a balloon floating through the trees.

Square World

The gravitational field on such a planet would still be approximately radial.

This means that as you walked from the centre of a face, it would feel as though you were walking increasingly up hill, causing you to lean further and further forward. At the edge, each face would appear to be at 45 degrees to your vertical and you would be standing face-to-face with any climber scaling the other face.

Water would flow 'downhill' to form circular pools at the centre of each face, as would the atmosphere, making any excursion to the edges of the world a very perilous venture.

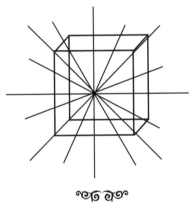

Star Gazing

Stars appear red or blue, not because they are moving, but because they are particularly cool or hot. There is a shift to the red or blue end of the spectrum caused by relative movement, but as Tom explained this would not cause the star to change colour. Astronomers know the shift has occurred because there are small gaps in a star's spectrum caused by elements absorbing very specific frequencies. These gaps come in identifiable patterns, and it is these patterns of gaps that are seen to move along the spectrum.

Before the Beginning

Before dealing with these two questions, we have to accept that there are some questions to which there are no answers because the question is nonsensical. For instance, "What would it be like to travel slower than stationary?" or "If I invented a universal solvent, what would I keep it in?" It may well be that the questions posed here fall into this category.

There is a general misconception about this theory. The Big Bang Theory does NOT say that all the matter in the Universe used to occupy an infinitely small space. Rather it says that the UNIVERSE itself used to occupy a single point. That Universe contained all the matter together with the familiar dimensions of length, breadth and height, and time. So time did not exist before the Big Bang, and three-dimensional space did not exist outside of it.

This is a very difficult thing for us to imagine, but it is what the theory says.

A Massive Problem

No. It is true that as objects become more massive, it becomes more difficult to change their direction and this must be true for the Earth also. The Earth would therefore need a greater force to keep it in orbit. The force is provided by the gravitational interaction between the Sun and the Earth, and this is proportional to the mass of the Earth. So, increasing the mass of the Earth automatically increases the gravitational pull on it by just the right amount.

Furthermore, it is generally accepted that, due to losses of lighter atmospheric gases (Hydrogen and Helium) there is a net loss of mass annually.

Pocket Money

Let us first assume that we play the game one hundred times. One condition is clear from the outset, the amount of coins in the gamblers' pockets cannot be under their control, as one of them would opt for zero coins to be a certain winner. Let us therefore assume, for simplicity's sake, that the choice is between 10 and 20 coins and that the allocation between the two players is decided at random, say by a computer, or the toss of a coin. It becomes evident that chances are equal and not favouring either player. This solution differs from widely held views, including that of Professor M. Kraitchik, who in his book Mathematical Recreations states that '...it is wise not to try to estimate the probability'.

୨୦୧ ୨୦୨

Ellipse

Assume for the sake argument that such a polygon existed, say a hexagon. If you were to circumscribe the hexagon with a circle then this circle would cut the ellipse in six places, i.e. the vertices.

As the illustration shows, an ellipse can only be cut in four places by a circle.

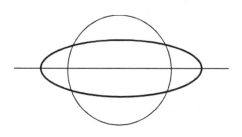

One Side of the Moon

No, it is not a coincidence. It is well known that the moon exerts tidal forces on the Earth's oceans. What is less well known is that there is also a tidal effect on the Earth's solid surface. The total result of this is to slow the rotation of the Earth, each day being slightly longer than the preceding one. This is why occasionally a "leap second" has to be added to the length of a day to keep clocks accurate. It is assumed that at some time in the past the moon was rotating faster, and that tidal forces in its surface slowed its rotation until it became synchronous. This phenomenon is quite common in the solar system; for instance, Titan takes 15 days and 23 hours to orbit Saturn and to rotate once about its own axis.

King of Cards

The bridge player must hold one Jack, one Queen, one King and one Ace.

Sum of All

a^2 and b^2 will always be larger than $2ab$. Independent of the relative values of a and b, $(a - b)^2$ will always be positive, but $(a - b)^2$ can be extended to $a^2 - 2ab + b^2$. Consequently $a^2 + b^2$ must be greater than $2ab$.

Gravity

There is no way to distinguish between 1) gravity and 2) acceleration. However, in a rotating spacecraft "gravity" will appear greater farther away from the centre of rotation, so there are several things that he could do:

1. Weighing an object with a spring balance will give a smaller reading near the ceiling.
2. A marble placed in the centre of the floor will roll towards the edge.
3. Two plumb lines will diverge rather than hang parallel.
4. He would not be able to spin a coin.

Zero Option

First split the figure into two triangles, ABC and ADC. Find the centroid of each triangle C1 and C2, to obtain lines C1C2. Then deal with triangles BCD and BAD to find C3 and C4, obtaining line C3C4. The intersection of the two lines is the centroid of the quadrilateral.

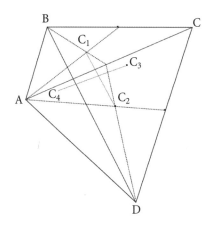

Pull of the Sun

Let us consider a simpler system first. Suppose there were only the Earth and the sun. If the Earth were stationary in space, the two would be drawn together by their mutual gravitational attraction. However, there is a more stable possibility: the sun's gravitational force is used to maintain the Earth's orbit rather than drawing it closer. The same process explains why the moon orbits the Earth. The sun's gravity is used to maintain the Earth/moon system in orbit. If all orbiting suddenly ceased, all the planets and their moons would instead fall towards the sun.

A Game of Three Sides

We know that any side of a triangle must be smaller than the sum of the other two sides.

This can be expressed as: $a < b + c$

It follows, then, that $a + a < a + b + c$

Simplified: $2a < a + b + c,$ *or*

$$a < \frac{1}{2} \ (a + b + c) = a < \frac{S}{2}$$

where S = the perimeter of the triangle.

Changing Sides

The original area = ab.

Let a be the side which is to be increased by $1/3$, making it $(a + \frac{a}{3})$ and b the side to be decreased by X. Then:

$$\left(a + \frac{a}{3}\right)(b - x) = ab$$

$$ab - ax + \frac{ab}{3} - \frac{ax}{3} = ab$$

$$b - x + \frac{b}{3} - \frac{x}{3} = b$$

$$\frac{4x}{3} = \frac{b}{3}$$

$$4x = b \qquad x = \frac{b}{4}$$

Four By Four

Let a be the side of a square, $(a + b)$ and $(a - b)$ the sides of a rectangle so that $2\left[\,(a + b) + (a - b)\,\right]$ equals $4a$, to make perimeters equal then:

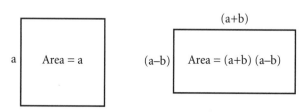

The area of the square is a^2, while the area of the rectangle is: $(a + b) \times (a - b)$ which equals: $a2 - b2$ which proves the point.

This example seems trivial but is the prelude to many interesting and thought-provoking problems.

Fred & Steve

As a general rule, there are four possible combinations with two children as far as their sex is concerned: Boy-Boy, Boy-Girl, Girl-Boy and Girl-Girl. It is important to realise that Boy-Girl and Girl-Boy are two different events.

Fred. As Girl-Girl is ruled out by premise, the probability that both are boys is 1/3.

Steve. Let us know consider the same four combinations in the order of older first and younger second. Boy-Boy and Boy-Girl are ruled out, making the probability of Girl-Girl 50-50.

On the Tiles

To fall at random and not to cross any line, the centre of the card must fall within a designated square whose sides are smaller by approximately 3.536 inches than the 4 inch and 8 inch square pattern respectively as shown in the illustration for the 4 inch square. This will allow the disk whose centre is within the designated square to clear any lines even if it falls so that its hypotenuse is perpendicular to the lines.

The illustration shows the 4 inch square and the disk (dotted lines).

The designated areas total 4.306 square inches for the 4 inch squares and 79.708 square inches for the 8 inch squares, making a total of 84.014 square inches.

Total area of floor unit is 576 square inches. Therefore the disk will not cross a line in approximately 1 in 6.856 throws, irrespective of how it lands.

The odds cannot be precisely computed because an irrational number, the square root of 2, is involved.

It is important to note that the odds against the disk not crossing the line given as 1 in 6.856 are for all cases when the diagonal of the disk falls at right angles to one of the lines. That is an extreme case and therefore in practice the odds are more favourable. These could be established, but the computation is rather complex. This problem could have been avoided by making the disk a circle instead of a square.

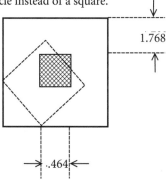

1.768

..464

Olbers' Paradox

Olbers himself provided the counter-argument.

1. While light from some of the galaxies we observe left them millions of years ago, the light of many millions of stars has not yet reached us. Our own galaxy is more than one hundred thousand light-years across.
2. The paradox assumes that all or most stars in the observable universe are alight at the same time. This is a misconception, because the life span of a star the size of our sun is limited to about 10 billion years. Although this is a long time, it is still finite.
3. Light may also be absorbed by dust.

Black to the Future

White is, strictly speaking, not a colour but the presence of all colours. A white bulb gives out all visible colours. Black, also, is not a colour but the complete absence of all light. By definition, therefore, there can be no such thing as a source of "black light".

Ringing Tone

There was a strip of railing nearby. Sound reflected back from each individual support rail arrived as a series of small echoes separated by a small interval of time. This caused a distinct ringing tone, the frequency of which was determined by the distance between the rails.

Sound Reasoning

Although light and sound are both wave forms, there are some significant differences, the major one being wavelengths. Sound waves are typically one meter in length, whereas the wavelength of light is about one ten-thousandth of a millimetre. Also, sound is "coherent" (continuous) whereas all light sources (except lasers) produce light in the form of millions of small packets, which are all jumbled up.

1. Waves will curl around obstacles (diffract) when their size is of the same order as the wavelength. Thus, sound will easily diffract around buildings, trees, etc. This can also be seen for light, but the obstacles have to be much smaller. For instance, a diffraction pattern can often be seen while looking at a bright point of light through a closely woven material.
2. Light can also be made to combine destructively, just like sound, but because of the small wavelength and lack of coherence this can be difficult to observe. This is what causes the characteristic "speckled" appearance of a spot of laser light and the colours seen in soap bubbles and films of oil.

Frequent Changing

There would be even worse problems if two or more transmitters attempted to broadcast the same program on the same frequency. As radio is a type of wave, at some points the two waves from the two transmitters could just happen to be synchronized so the radio would pick up a very strong signal. At other places the waves could be completely out of synch and would cancel each other out. Thus, as you moved around, the radio would continually go from extremely loud to no signal at all within the distance of a few meters.

Light Fantastic

In one of the rooms, the light is provided by sodium lights, which provide a very pure yellow light. In the other room the light is provided by a mixture of red and green lights. Red and green lights, when mixed, appear yellow. In the first room, both red and green paint appear black, while in the second room the red and green paints reflect their own colours.

︎

Faster Than Light

My friend correctly identified three situations in which something travels faster than light. However, none of them contradicts Einstein, whose work could be more correctly summarized as: "The maximum speed at which information can travel is represented by the speed of light in a vacuum." In the first two examples, nothing physical moves faster than light the barber's pole stripes only appear to move along the pole. The spot of light at one end of the sweep is different from the one at the other end as it is made up of different photons. Neither system could be used to carry information. The third example has been observed experimentally. Such particles break the "light barrier" and produce minute photonic flashes equivalent to sonic booms.

Life in Technicolour

Part of the mysterious beauty of the rainbow:
1. You can see the spectrum only if the angle of refraction between the sun, the drop of water, and your line of vision is between 40° and 42°.
2. Lunar rainbows are also possible but rare, because moonlight is not as strong as sunlight and its intensity varies with the phases of the moon.
3. This phenomenon is called dew-bow and is caused by the water drops on the grass.

Cold Feet

The two floors were probably at exactly the same temperature. However, carpet is a much poorer conductor of heat than are the tiles. As the tiles were taking heat from my feet faster, the tiling felt colder than the carpet.

Car Headlights

It would work. However, it was not introduced because the disadvantages would outweigh the benefits.

1. Not being able to see the headlights of approaching cars would be dangerous, particularly when visibility is reduced during heavy rain or fog.
2. The windshield's polarization would absorb a fair amount of light in the street scene, reducing visibility generally.
3. At any rate, it would be difficult to provide an effective filter for windshields in view of their irregular shapes.
4. Coloured stress patterns would become visible in the wind-shield and this would be distracting for the driver.

Vision and Sound

With two ears you can identify the direction from which a sound is coming. The same principle is used in direction-finding equipment to locate the source of illegal radio transmissions.With only one eye you would be unable to see objects in three dimensions, nor would you be able to estimate distances accurately.

Toy Boat

As the steam enters the cooler tubes it condenses, creating a vacuum that will draw in water. However, such water as is sucked in will not do so in the form of a jet, but will come from all directions and will not therefore have the same propulsive force.

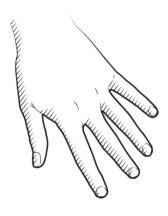

Breath on Your Hand

There are two possible explanations involving: a) evaporation and b) expansion. The blown air moves more rapidly, causing moisture in the skin to evaporate quickly, which produces a cooling effect. As the air passes through pursed lips, it expands and cools. This is the opposite effect to air being compressed in a bicycle pump becoming warm. You can prove that both effects are present with a piece of clear food-wrap. If you place this over your hand, preventing evaporation, the blown air still feels slightly cold. Wetting the outside of the food wrap allows evaporation to occur, restoring the full cooling effect.

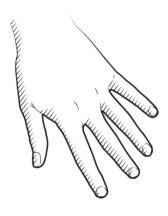

The Foehn

There are three questions to be answered:

1. Why is the wind warm? The meteorological conditions have to be right for the cold mountain air to sink in the form of a fast-moving wind. As the air sinks it encounters the higher pressures found at lower altitudes and is compressed. This compression causes warming; the result of a phenomenon known as adiabatic process.
2. Why is the wind dry? The air was originally cold and cold air can hold less moisture than hot air.
3. Why does it have psychological effects? Nobody is sure that it does, or if so what the reason might be.

Drinking Bird

The construction of this gadget is as complex as it is ingenious. The base is partly filled with a suitable vapour-forming liquid (L) and connected to the head by the tube (T). As T is immersed in L, it produces vapour in the base as well as in the tube and the head. Because the immersion of T in L forms a seal, the two vapour systems are not connected. One should also mention that the centre of gravity (COG) of the contraption is below the fulcrum, so that its position is, to begin with, stable, as illustrated.Now you wet the top of the bird. As water evaporates, the vapour in the head cools, lowering the pressure. The higher pressure of vapour in the base forces liquid up the tube, relocating the COG until the bird's head tilts forward, dunking its head in the water. In doing so, T is lifted out of L and the two pockets of vapour connect in the tube and equalize pressure. This, together with the buoyancy created by immersing the head, moves the COG again to below the fulcrum. The bird rights itself and the whole cycle begins again.What makes this toy interesting is in considering where the energy comes from to keep it going. The answer is that the energy comes from the heat in the water – which gets progressively colder.

TV Interference

Your TV antenna receives two signals, one direct and one, slightly later, reflected from the plane. These television waves interfere with each other, sometimes constructively and sometimes destructively, depending on the position of the plane. This would cause the picture to flicker in a disconcerting manner.

Invisible Man

Unlike sound, natural light is not coherent, that is to say, it consists of jumbled packets, making it difficult to produce another beam, which is the exact opposite. (An exception to this is laser light, which is the only source of coherent light). If destructive interference were somehow achieved, the effct would be a black silhouette, not an invisible man. The real challenge would be how to enable rays of light to pass through solid matter without deviation (even glass refracts light). Therefore the invisible man will remain a figment of the imagination.

Red Eye

In low light conditions, the pupils of our eyes are open wide, to capture as much light as possible. Most popular cameras have built-in flashes situated near the lens and this flash causes a large amount of light to enter the eye, the reflection from the fundus at the backof the eye manifests in the picture as 'red eye'. Professionals either move the flash higher, so the light entering the eye misses the retina, or they do not fire the flash directly at the subject, but bounce it off the ceiling or a reflector. Most cameras have "anti-Red-Eye" devices, which reduce the effect by shining a bright light towards the subjects a short while before the picture is taken, thereby narrowing the pupil.

Many vertibrates including some apes (although not humans) have an additional reflective layer (the tapetum) behind the retina so that the light sensitive cells have an extra chance of absorbing the photons of light. This reflective layer results in the phenomenon of 'eye shine' in many nocturnal animals.

Looking Blank

Our eyes do not produce an absolute picture, our brain also interprets what we see. Our eyes compare brightnesses of nearby areas. For instance, if you look at a street light against a twilight sky, you will 'see' that the sky near the lamp appears darker than elsewhere. Similarly, the grey areas near the bright part of the TV picture will be interpreted as being black.

Enlightenment

The candlewick is not burning; if it did, the candle would not last for very long. The solid wax also does not burn. The heat from the flame first melts the wax, which is drawn up the wick where it is vaporised by the heat. The vapour combines chemically with the oxygen in the air in a process we call burning, thus providing the heat needed to keep the process going. Particles of carbon are produced which glow incandescently, producing the light. The hottest part is therefore on the outside of the flame where the vapours and the air are mixing most efficiently. This is why the flame is brighter on the outside.

Ghosting

I was looking though a double glazed window. Most of the light came straight through, but some was reflected from the inner pane, and then the outer pane and back to me. The produced a faint second image that I was able to see because of the dark wall.

Travelling Light

Light travels at a very high speed. It could travel round the world seven times in one second. Over the distances that Galileo was using, we could consider the light travelled instantaneously. What Galileo was really measuring was his own and his assistant's reaction times, and this would be independent of distance. The answer is therefore a).

Looking Glass

There are several possible solutions. For instance, you could bounce the image from a projector off a mirror onto a screen. What you see on the screen is an image of the film in the projector. What you see in the mirror is an image of that image. Alternatively, think of a concave mirror, such as a shaving or make-up mirror. If you stand some way away from it and look at the reflection of a distant object, it will appear to be upside down, AND in front of the mirror. You can prove that the image is 'on our side' of the mirror, by projecting it onto a piece of white paper.

Flying in a Circle

The mirror would have to rotate at half the speed of the bird.

The image of the bird will always be as far behind the mirror as the bird is in front, this means that the mirror always bisects the angle, x, between the bird and its image. When the bird moves through a certain angle, y, the angle x decreases by y. The mirror must rotate through half that angle, y, so that the image remains at the same point and to bisect the new angle.

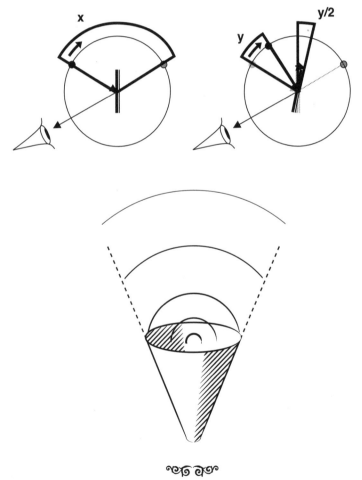

The Wine Glass

All objects have a natural frequency at which they will vibrate if given energy. You can hear this note if you hold the wineglass by the stem and gently flick the rim. For a wineglass this is several hundred Hertz (cycles per second). If you rub a dry finger across glass, it bumps unevenly along as friction makes it stick and then releases it. The frequency of this will be a few tens of Hertz. Wetting the finger reduces friction so that its frequency is close to that of the glass. Providing the finger is moving at the correct speed, a phenomenon called resonance occurs. The glass vibrating under the finger nudges it at exactly the right moment so that the finger and the glass have EXACTLY the same frequency. At this point vibration of the glass increases dramatically, making it sing.

Take A Bow

The surprising answer is that bowing is actually also plucking. The bow is primed with a resin called colophony, which makes the bow stick to the strings momentarily, until the player forces the bow forward thus effectively plucking the string. The pluck and slide sequence translates into the ear as a continuous smooth sound.

String Theory

The frequency of wind instruments such as flutes, bassoons and clarinets depends on the speed of sound. As these instruments are warmed by breathing into them, the speed of sound and therefore the pitch increases. As to string instruments, warming of the strings by friction will expand them, thus lowering the pitch.

Snow Landscape

The effect works best with loosely packed snow. There are many air gaps into which the sound penetrates, and moving flakes dissipate(s) the energy. The sound is therefore absorbed rather than being reflected. The small perforations in an acoustic tile work in the same fashion.

Frequencies

The sound that hits your ear when you talk normally comes partly through the sound waves and partly by conduction through bone. This applies largely to the low frequency range which make your voice sound richer to yourself, though others hear you as you perceive your recorded voice.

Dungeons of Dionysius

Yes it works. If a room is constructed in the shape of an ellipsoid then any sound emitted from one focus will be clearly heard at the other focus as long as the prisoners are positioned at one focus and Dionysius was at another, within the same cell.

Soup on the Rocks

The hot soup warms the outer layer of the cube first. Before melting, this warmed ice changes in volume, distorting the crystal lattice of the ice. This causes fractures to suddenly appear, which makes the sound. We say that thermal stresses have been produced in the ice cube.

The Blowing Horn

While the frequency of the horn is indeed a constant, there is an apparent change in the observed frequency. As the source of sound approaches, more sound waves per second impinge on the observer. Conversely as the car recedes fewer sound waves per second reach the observer. The phenomenon is called The Doppler Effect after Christian Doppler (1803-1853), the Austrian physicist. His discovery is of immense importance in astronomy, inasmuch as the Doppler Effect also applies to light waves. The frequency of light waves is extremely high, of the order of several hundred billion waves per second, nevertheless a spectrum shift to the red can be observed, for stars which move away from us.

Walls Have Ears

Unfortunately not. Firstly, the amount of energy involved is very small – it is just that our ears are very sensitive. You probably could not collect enough energy to run a small calculator, let alone a light bulb or heater. Secondly, a microphone absorbs only a small proportion of the sound energy that lands on it, so there would be plenty left over, to cause you annoyance.

Speed of Sound

Think of one of the wheels. The centre is connected to the axle, and travels at the same speed as the car. The part of the wheel in contact with the ground is stationary. The top of the wheel has a forward motion relative to the car.

This means that the top of a wheel always has a forward speed twice that of the car; so at 380 MPH this part is travelling 760 MHP, or faster than the speed of sound. This could cause shock waves that could destabilise the vehicle. Therefore the wheels of such high-speed cars are always enclosed, so that a pocket of air is carried along with the wheel. This means that although the top of the car is travelling faster than sound relative to the ground, it is not travelling faster than sound through the air that surrounds it, therefore it does not 'break the sound barrier'.

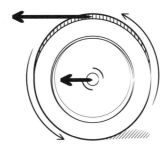

Tidal Power

No. Water would not flow along the pipe. Although the depth of water is different at the two ends, the weight of water above the ends of the pipe, and hence the pressure is the same. This is because the Moon's gravitational effect cancels some of the Earth's gravity in the high tide area. Therefore the two different depths of water cause the same pressure.

Super Computing

Until a few years ago, most people would probably have said "yes". However, modern work with Chaos Theory suggests otherwise. Let us take a situation much simpler than the horse race. Suppose we have a pendulum with an iron bob, free to swing near two magnets; one painted red and the other green. If set swinging, the pendulum will move in a complex way before the bob comes to rest over one of the magnets. We wish to be able to predict, from a given starting point which magnet will be the final resting-place. So I determine a starting point, measured let's say to the nearest centimetre, let the bob swing and then on graph paper I colour a square according to which magnet it selects. I then repeat for all the squares.

We would end up with a pattern of red and green squares. Suppose we wished to improve the accuracy, so I selected one of the squares, perhaps a green one, and measure 100 sub-positions to the nearest millimetre. We would find that although the centimetre square was green, the millimetre squares would NOT all necessarily be green. There would be a further pattern of red and green squares. We could continue refining this process ad infinitum, discovering new red and green squares as we went. In other words the predicted outcome depends entirely on the accuracy with which we make the original measurements. As 100% accuracy is not possible, we can never predict the outcome. This must apply to the result of the horse race also.

City Speed Limit

The subject was, of course, Einstein's Relativity Theory. Time is the rate at which things happen; for instance how fast your heart beats, radioactive atoms decay, pendulums swing, objects fall. Einstein hypothesised, and subsequent experiments proved, that time does not always run at the same rate. For instance, when moving, one's time slows relative to a stationary observer. The reason why you and I have never noticed it, is that one would have to travel an appreciable proportion of the speed of light for it to become apparent. The young man in the story worked on the railways and spent an appreciable time travelling. His personal rate of time was therefore less than his granddaughter who did not travel so much. So she aged much faster, eventually overtaking him in age.

Dark Matter

It is not the 'solidness' of matter that prevents objects passing through one another. On the outside of all atoms and molecules there are clouds of negatively charged electrons. When objects come into close contact, there is a large force of electrostatic repulsion between the two surfaces and it is this that prevents the penetration. If an object had no charge, it could theoretically pass through solid matter. There are no large objects like this, but there are small particles that are neutral and these have great penetrating power. For instance, millions of neutrinos, born in the nuclear maelstrom of our Sun are passing harmlessly through your body every second, they continue through the Earth as if it were not there. It has been estimated that to stand an even chance of stopping a neutrino you would need a lead block whose thickness was ten times the distance from here to the nearest star.

Revolutions

No. Reducing the load would suddenly allow the generators to run more freely, and there is a danger that they would rotate too fast, but there are fail safe devices to prevent this causing damage. After all, the same effect may be caused when a district suffers a blackout, and this has never yet caused any damage to a power station.

Turning on the Power

Yes, but probably impractical. The amount of energy needed to overcome the friction in the bearings would probably be greater than the energy generated by the slowly turning gyroscope. The energy generated would come from the rotating Earth that would slow slightly as a result. This is the same energy that is tapped by tidal barriers to generate electricity, which is probably a far more efficient way of doing it.

Clocking Off

There would be all sorts of unfortunate effects. The further North, or South, you lived, the longer your working day would be in the Summer. Those living beyond the Arctic or Antarctic Circles would have a six month working day. Eggs would take much longer to boil in Winter than Summer. And what about speed limits? Thirty miles an hour would be much faster during the short Winter hours, unless we redefined length also. If distances were also shorter in Winter, it would be difficult calculating the amount of fuel needed for a journey, unless volume was also varied. Which would mean that fuel would be cheaper at certain times of the year, unless we also changed the value of money.

The Transporter

There is probably no definitive answer to this question, which has more to do with philosophy than science. One could argue that there is a completely natural equivalent. Your body is continually replacing itself, so that after a number of years there are very few of your original molecules left, and yet you are the same person. This is a little like the old joke about the axe at the Tower of London, which is 400-years-old but still looks in good condition because during that time it has had three new heads and six new handles.

Perhaps the key is to ask yourself if, having had this problem posed to you, you would be prepared to enter the transporter, knowing that your consciousness was about to end and another identical person with all your memories would be created at the other end. If you answer 'Yes' then for you, there is no dilemma. If you are convinced that the same person emerges at the other end, then you might like to ask yourself what would happen if the Reassembler made a second person. Which would be the real traveller?

Ultimate Weapon

It would not work, even theoretically. In the first instance the major devastation would occur in the country resorting to such a weapon although, according to David Stone's theory, there would be damage in the USA provided the jumps were timed in regular intervals to amplify the resulting ground waves. Stone estimates, however, that a jump would have to take place about once every hour for some considerable time. So, apart from the devastation at home, the attacking country's economy would certainly suffer irreparable damage. Theoretically at least, the USA might indeed consider inviting all NATO members to organize counter-jumps, carefully timed, to cancel the ground waves initiated by their adversary as an alternative to using advanced weaponry in defence of their shores and avoid the possibility of starting a world conflagration.

Charging for a Flight

No. It is true that a voltage would be generated between the wing-tips as the plane flew through the Earth's magnetic field. However, to make use of it we would have to complete the circuit, and that would require running wires to the ends of the wings. Those wires would also be traveling through the Earth's field and would also generate the same voltage. It would be like connecting two identical batteries back to back: the two effects would cancel each other out. If we consider other forms of transport, we might be more successful. By a similar process a train would generate a voltage which could be picked up from the rails by contacts lying beside the track. I calculated that a train traveling at 100 mph (160 kmph) would generate about 1.5 millivolts. That is obviously not enough to solve any energy crisis! NASA is experimenting to see if this is a viable power source for spacecraft. A long wire, several miles long is trailed through the Earth's magnetic field. The electric charge is dissipated from sharp points at both ends of the wire.

The Scientist

2, 3, 10 and 10

Keeping in mind the third and fourth hint, and to maintain the difference of eight, the first and third cards can be 2 and 10 or 10 and 2. But keeping in mind the first hint, we know that the first card is 2 and the third is 10. Similarly, if we see the fifth hint, the difference between the second and fourth card should be 7. Thus we know that the second card is 3 and the fourth card is 10.

Time Bomb

The design would not work. First, there is no such thing as a perfect reflector of sound. Second, the amount of energy in a watch is quite small. But even without these two objections the proposal would still be impractical. Air is a very inefficient transmitter of sound and consequently sound energy is continually being converted into heat. The energy thus escapes from the box in the form of heat.

༄ༀ ༀ༄

The Rally

Assume that instead of the return trip, another motor car is starting from Lugano at the same time as Andrew leaves Basle, both travelling the same route. It is logical that the two cars must meet somewhere along the way, irrespective of the speed at which they are travelling. Another proof can be obtained by plotting Andrew's progress on both journeys, using time and distance as co-ordinates as shown in the graph below. If the average speed of each leg is recorded, the spot which Andrew passed at the same time of day on both journeys could be ascertained.

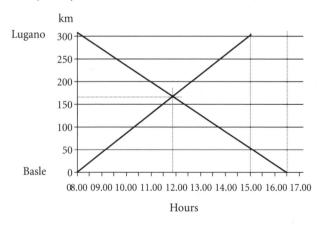

༄ༀ ༀ༄

Three Cities

Taking the earth's circumference as 25,000 miles, cities A and B are anywhere from zero to 7,000 miles apart, as the greatest distance between the three cities must lie on a great circle, approximately equal to the circumference measured at the equator.

Piggy Bank

Using an idea first developed by Friedrich Gauss (1777-1855), the German mathematician and scientist, when he was eight years old, we can argue as follows:

The series $1 + 2 + 3 + 4 + \dots N$ has an intriguing feature:

$$1 + N, 2 + (N - 1), 3 + (N - 2)\dots$$

add up to the same sum. For instance, dealing with:

$$1 + 2 + 3 + 4 + 5 + 6 + 7 + 8 + 9 + 10$$

you find that the sum of first and last, second and one but last... always adds up to eleven. If you continue five times, covering all digits, you obtain the sum of the series, i.e. 55.

As a general formula,

$$(1 + N)\,\frac{N}{2}$$

will deal with our problem:

Let X be my present age. Then,

$$(1 + X)\,\frac{X}{2} = 300$$

Simplify:

$$X^2 + X = 600$$

Solve quadratic equation using standard procedure:

$$X^2 + X + \frac{1}{4} = 600 + \frac{1}{4}$$

$$(X + \frac{1}{2})^2 = \frac{2401}{4}$$

$$X + \frac{1}{2} = + \frac{49}{2}$$

$$X + 24$$

The Price is Right

In economic terms the supply of the Mona Lisa is perfectly inelastic, as irrespective of the price bid, there is only one painting for sale. The demand curve represents the diminishing number of bidders as the bid price increases. At price P there is only one bidder left, and therefore the demand curve stops.

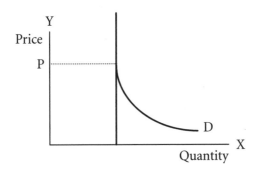

Pond Life

Take a net and catch say a hundred fishes from the pond. Dab each of them with a spot of paint, which will not dissolve in water, and return them to different parts of the pond. Wait until they disperse and then net once again another hundred. Some from the second catch will have the blob.Let us say that six fish from the second batch are marked. It is reasonable to assume that the first catch also represented six per cent of the fish population, which makes the total just short of 1700, proving that your friend's estimate was not badly off the mark. If you repeat the exercise once or twice your result will become more accurate.

⁂

Combination Lock

The response to my suggestion that the combination might be 4, 1, 5, 3, 2 proves that the two digits in the correct position must be adjacent, otherwise a digit would follow one which is in the correct position. This would mean three digits correctly located, which is not the case. Therefore 4–1, 1–5, 5–3 or 3–2 are in the right position. The pairs 1–5 and 5–3 must be eliminated as a second following digit could not occur. If 4–1 is correctly located, the combination would have to be either 4, 1, 2, 5, 3 or 4, 1, 3, 2, 5. These solutions must be eliminated because of the response to my first guess. Consequently 3 and 2 are in the correct position. The combination 1, 5, 4, 3, 2 also contradicts the first comment, leaving 5, 4, 1, 3, 2 as the only possible solution.

⁂

Imitation of Life

The art dealer started with 62 paintings. To solve the equation becomes unwieldy, even for two customers. Let X be the number of paintings in stock. Then:

$$X - \left(\frac{X}{2} + 1 \right) - \left\{ \frac{1}{2} \left[X - \left(\frac{X}{2} + 1 \right) \right] + 1 \right\} = 0$$

making $X = 6$.

Logic yelds a quick solution. Start with the fifth buyer who receives half of the paintings left plus one. To result in zero left, the dealer must have had 2 paintings when serving him. Doubling up and adding 2, i.e. 6 must have been the stock when serving the fourth buyer.

A general formula for N buyers would be:

$$X = 2^{(N+1)} - 2$$

whereby X = commencing stock.

The Manuscript

No. He could only achieve the daily average of 4,000 words if he were to complete the second half of his manuscript in no time. If you don't believe me, think again.

Centroid of a Triangle

Using the zero option principle, imagine the triangle to consist of an infinite number of strips with zero width, all parallel to AB. Each strip will have its centre of gravity in midpoint. The locus of all these points is the line from C to the midpoint of AB. Repeat the procedure from the vertices A and B making M the centre of gravity of the triangle. The proof that the three lines concur is inherent in the fact that a triangle or indeed any object can have only one centre of gravity.

The Unwilling Triangle

Pythagoras' theorem which applies to all right-angled triangles states:

$$a^2 + b^2 = c^2$$

where a and b are the sides and c is the hypotenuse. In our case a = b, and therefore the formula is reduced to:

$$c^2 = 2a^2 \text{ or } c = a \sqrt{2}$$

The right side is irrational and thus can never be a whole number.

Fibonacci's Sequence

Let x and y be the first two terms of a Fibonacci sequence. Then the sequence would read:

x, y, x + y, x + 2y, 2x + 3y, 3x + 5y, 5x + 8y, 8x + 13y, 13x + 21y, 21x + 34y

The sum of the first ten terms is equal to 55x + 88y which is eleven times the seventh term 5x + 8y.

It Must Be Even

A number is even if it has at least one even factor. Consider a number raised to a power, x^y. There are two possibilities:

a) x is even, in which case whatever the value of y, x^y must also be even. As the difference between two consecutive even numbers is 2, then the difference between any two even numbers will be even and the sum of two even numbers will also be even.

Therefore $x^a - x^b$ is even, and $x^c + x^d$ is even

b) x is odd, in which case whatever the value of y, x^y must also be odd. As the difference between two consecutive odd numbers is also 2, then the difference between any two odd numbers will be even, and the sum of two odd numbers will be even also.

Therefore $xa - xb$ is even, and $xc + xd$ is even.

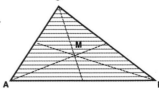

Discovery

We know that $a^2 - b^2$ can be expressed as $(a + b)(a - b)$. By premise $(a + b)(a - b)$ should be equal to $(a + b)$. This is only possible if $(a - b) = 1$.

Perfect Time

Let S_1 be the speed of HH, and S_2 the speed of MH. We know that speed equals distance divided by time. Consequently:

$$S_1 = \frac{1}{12} \text{ of } \frac{3\pi}{2} \ p = \frac{3\pi}{24} = \frac{\pi}{8}, \text{ and}$$

$$S_2 = 2\pi$$

Therefore:

$$S_2 = 16\,S_1$$

Using simple logic, the solution is still simpler:

If both hands are the same length, the minute hand must travel 12 times faster than the hour hand. The speed of the end is proportional to the length of the hand. Therefore the relative speed of the minute hand

$$= \ 12\,\text{x} \,—\, = 16.$$

Unusual Spheres

We know the surface area of a sphere to be $4\pi r^2$, and the volume to be:

$$\frac{4}{3}\pi r^2$$

According to the specification, both $4r^2$ and $-r^3$ must lie between 1000 and 9999.

To meet these limits, r must be larger than 15 and smaller than 20. In other words, r must lie within the range 16 to 19.

But for $-r^3$ to be an integer, r must be divisible by 3. Only 18 in the prescribed range meets this condition, making the area 1296ϖ, and the volume 7776ϖ.

❧ ☙

The Billiard Table

Extend AX to point E. Angle AXC is clearly equal to angle DXE. Also XP = XE. As AE is a straight line it follows that AXP is the shortest possible rebound.

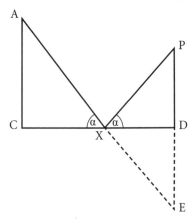

❧ ☙

Bullseyes

Assuming that you have no special aiming skill, the probability of scoring 90, 42 or 27 must be proportionate to the target area, which for the bull's-eye = $\pi r^2 = 9\pi$. We can ignore π and therefore the area of the 12 inch circle = (36-9) = 27, and for the 18 inch circle = (81-36) = 45. Multiply the areas with the appropriate score:

> 9. = 810
>
> 27. = 1134
>
> 45. = 1215

Total 81. Divide each score by the total area to get 10, 14 and 15. Adding up and multiplying by 10 gives you 390 as total probable score.

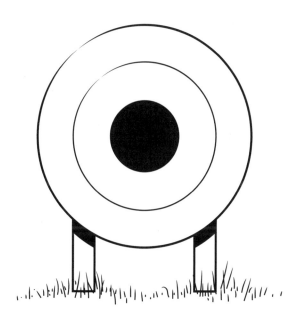

PUZZLE NOTES